ALI THE FLYING TIGER

Awesome Book of Universe for Kids

A Fast Journey of the Universe

UNIVERSE BOOK SERIES

Created by Ali Naeem

ABOUT AUTHOR

Ali Naeem is Software Engineer, Ethical Hacker, Graphic Designer & Children's Book Author. He is also the founder of "Ali The Flying Tiger" Company.

Contact

AliTheFlyingTiger1@gmail.com

01

What is the Universe?

Visit for More Books
www.alitheflyingtiger.com

The universe is everything. It contains everything that exists including the Earth, Planets, Space, Sun, Moon, Stars, and Galaxies.

04

02

WHAT IS ASTRONOMY?

Visit for More Books
www.alitheflyingtiger.com

Astronomy is the study of the Universe and everything in it, including Planets, Stars, Galaxies, Comets, and Black Holes.

03

What is the Big Bang?

Visit for More Books
www.alitheflyingtiger.com

Scientists Believe the Big Bang may have created the universe, 13.7 Billion Years ago. The Big Bang is the main theory that explains the origins of the universe.

04

WHAT IS GRAVITY?

Visit for More Books
www.alitheflyingtiger.com

Gravity is an incredible invisible force that pulls objects toward each other.

10

05

WHAT ARE STARS

Visit for More Books
www.alitheflyingtiger.com

Stars are Huge, Glowing balls of Gases.
Stars are made up mostly of hydrogen and helium.
Our Sun is also Star.

12

06

What is the Solar System?

Visit for More Books
www.alitheflyingtiger.com

The Solar System is made up of the Sun, Moon, Earth, Other Planets, and all of the smaller objects that move around it.

14

07

WHAT IS THE SUN?

Visit for More Books
www.alitheflyingtiger.com

The Sun is the only star in the Solar System It's at the center of the Solar System. It is a huge, spinning, glowing sphere of hot gas.

16

08

What is the Moon?

Visit for More Books
www.alitheflyingtiger.com

The Moon is the 2nd brightest object in the sky after the Sun. The Moon travels around the Earth in a circle called an orbit.

18

09

What are Planets?

Visit for More Books
www.alitheflyingtiger.com

Planets are the largest objects in the solar system after the Sun.

20

10
Names of Solar System Planets

Visit for More Books
www.alitheflyingtiger.com

The eight planets are Mercury, Venus, Earth, Mars, Jupiter, Saturn, Uranus, and Neptune

11

What are Exoplanets?

Visit for More Books
www.alitheflyingtiger.com

Planets that orbit around other stars are called exoplanets.

24

12

What are Dwarf Planets?

Visit for More Books
www.alitheflyingtiger.com

25

Smaller Planets are called Dwarf Planets.

13

What is a Galaxy?

Visit for More Books
www.alitheflyingtiger.com

A galaxy is a group of stars and other space stuff.

14
Types of Galaxies?

Visit for More Books
www.alitheflyingtiger.com

There are 3 types of galaxies - Elliptical, Spiral, and Irregular.

30

15

What is the Milky Way?

Visit for More Books
www.alitheflyingtiger.com

The Milky Way Galaxy is our Earth's Galaxy.
It is a spiral galaxy.
It is estimated to contain approximately 100 billion stars.

16

WHAT ARE METEOROIDS?

Visit for More Books
www.alitheflyingtiger.com

A meteoroid is a chunk of metal or rock which travels through space toward the earth is called a meteoroid.

34

17

What are Meteorites?

Visit for More Books
www.alitheflyingtiger.com

Meteorites are rocks of space, which are found on Earth after they fall from the sky.

36

18

WHAT ARE METEORS?

Visit for More Books
www.alitheflyingtiger.com

Meteors are also called shooting stars. Meteors are meteoroids that burn in the Earth's atmosphere.

19

WHAT ARE COMETS?

Visit for More Books
www.alitheflyingtiger.com

Comets are cosmic snowballs of frozen gases, rock, and dust that orbit the Sun.

40

20

WHAT ARE ASTEROIDS?

Visit for More Books
www.alitheflyingtiger.com

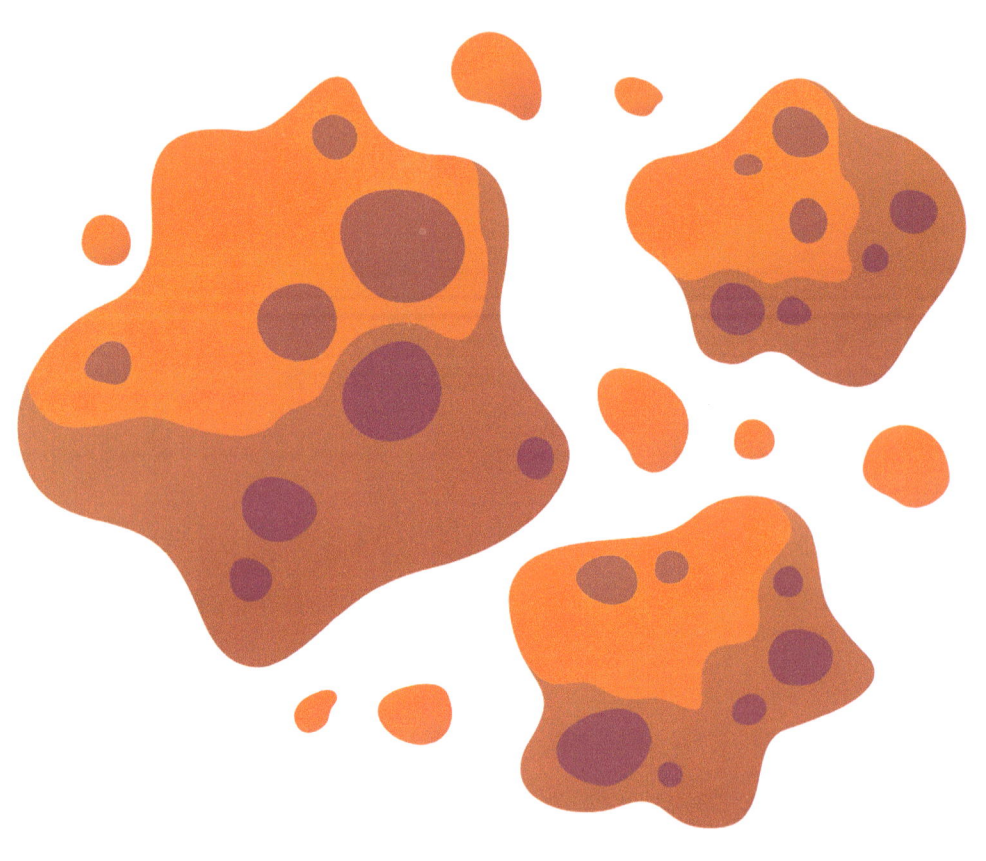

Asteroids are small, rocky objects that orbit the Sun.

42

21

What is a Nebula?

Visit for More Books
www.alitheflyingtiger.com

43

A nebula is a cloud of gas and dust in space. Nebulas appear in many shapes and colors.

22
What Is a Supernova?

Visit for More Books
www.alitheflyingtiger.com

45

A supernova is the explosion of a star.

46

23

What is Black Hole?

Visit for More Books
www.alitheflyingtiger.com

47

Black holes are the strangest objects in the Universe.
A black hole is an area in space with an incredibly strong force called gravity.

48

24

What are Quasars?

Visit for More Books
www.alitheflyingtiger.com

49

Quasars are the brightest objects in the universe.

50

25

What is Dark Energy?

Visit for More Books
www.alitheflyingtiger.com

Dark energy is something that scientists believe fills all space. It turns out that "empty space" is more than just nothing.

52

26

What is Dark Matter?

Visit for More Books
www.alitheflyingtiger.com

Dark matter is an invisible type of matter theorized to make up the majority of all matter in the Universe.

54

27

Death of a Star

Visit for More Books
www.alitheflyingtiger.com

All stars eventually run out of their hydrogen gas fuel and die.

28

What is Light Year?

Visit for More Books
www.alitheflyingtiger.com

A light-year is the distance light travels in one Earth year.

29

WHAT IS THE SPACESHIP?

Visit for More Books
www.alitheflyingtiger.com

Spaceship is a vehicle used for space travel.

30

What is a Satellite?

Visit for More Books
www.alitheflyingtiger.com

A satellite is an object in space that orbits or circles around a bigger object.

Copyright Notice

Subject to the provisions of this notice, this Book and all its content, information, or material is the copyright of Ali The Flying Tiger, together with its licensors. Accordingly, your use of our Book or its services does not constitute any license to use the copyright in our Book.

Except to the extent permitted by the applicable copyright law or Ali The Flying Tiger, any form of use, reproduction, or redistribution of part of all of the content, information, or material on this Book in any form is hereby prohibited.

© 2022 Ali The Flying Tiger. All rights reserved.

www.ingramcontent.com/pod-product-compliance
Lightning Source LLC
Chambersburg PA
CBHW041932240526
45473CB00034B/929